Carrots

by Gail Saunders-Smith

Pebble Books
an imprint of Capstone Press

Pebble Books

Pebble Books are published by Capstone Press
818 North Willow Street, Mankato, Minnesota 56001
http://www.capstone-press.com/

Library of Congress Cataloging-in-Publication Data
Saunders-Smith, Gail.
Carrots/by Gail Saunders-Smith.
p.cm.
Includes bibliographical references (p. 23) and index.
Summary: In simple text and photographs describes the growing cycle of the carrot, from planting through cooking and eating.
ISBN 1-56065-488-0
1. Carrots--Juvenile literature. [1. Carrots.] I. Title.

SB351.C3S28 1997
641.3'513--dc21

97-23586
CIP
AC

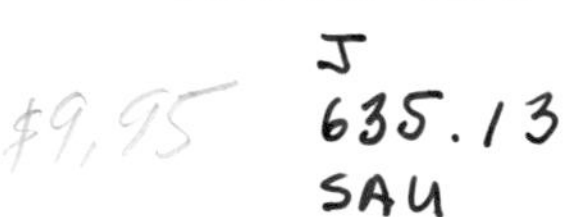

Editorial Credits
Lois Wallentine, editor; James Franklin and Timothy Halldin, design; Michelle L. Norstad, photo research

Photo Credits
Michelle Coughlan, 18
Winston Fraser, cover, 3, 4
Dwight Kuhn, 6, 8
Michelle L. Norstad, 14, 16, 20
Unicorn Stock/Shellie Nelson, 10; Jeff Greenberg, 1, 3, 12

Table of Contents

We plant carrots
in the garden.

We water carrots
in the garden.

We weed carrots
in the garden.

We pull carrots
in the garden.

We wash carrots
in the garden.

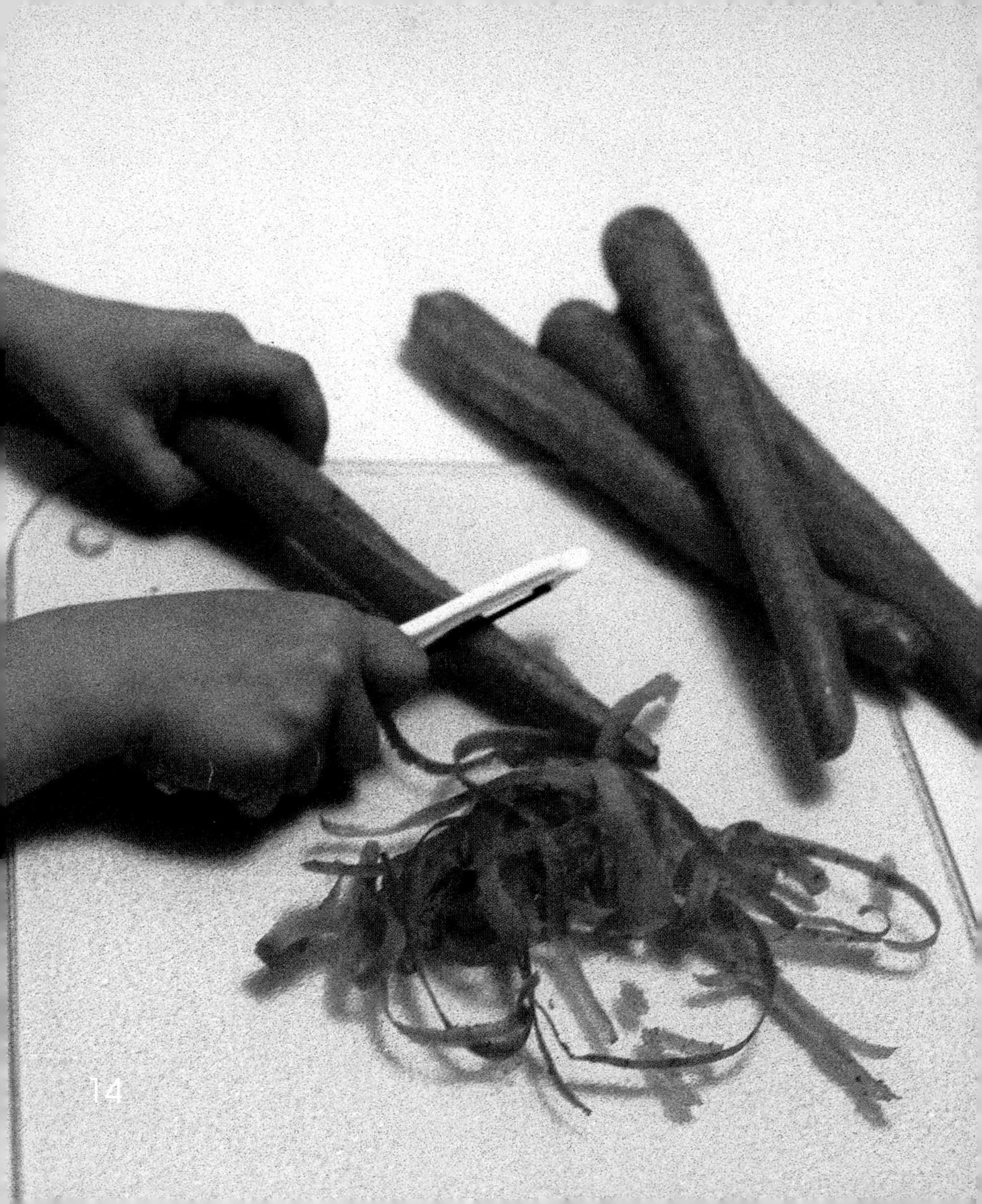

We peel carrots
in the kitchen.

We slice carrots
in the kitchen.

We cook carrots
in the kitchen.

We eat carrots.

Words to Know

cook—to prepare and heat food

peel—to remove the skin of a vegetable or fruit

plant—to put a seed in the ground so it can grow

slice—to cut into thin pieces

water—to pour water on something

weed—to pull unwanted plants from a garden

Read More

Gibbons, Gail. *From Seed to Plant.* New York: Holiday House, 1991.

Jennings, Terry. *Seeds.* New York: Gloucester Press, 1988.

Wexler, Jerome. *Flowers, Fruits, Seeds.* New York: Prentice-Hall Books for Young Readers, 1987.

Internet Sites

Calvin's Carrot Page
http://www.dole5aday.com/about/carrots/carrots.html

Carrots
http://uaexsun.uaex.arknet.edu/Vegfacs/carrots.html

Carrots from Carrot Paradise!
http://www.taylorcarrots.com/index.htm

Note to Parents and Teachers

This book describes the process of growing carrots from planting through eating. The noun is in its plural form. This provides the opportunity to discuss what plural nouns are and how to form them. The verbs change on each page. The photographs clearly illustrate the text and support the child in making meaning from the words. Children may need assistance in using the Table of Contents, Words to Know, Read More, Internet Sites, and Index/Word List sections of the book.

Index/Word List

Word Count: 51

FREE PUBLIC LIBRARY UNION, NEW JERSEY
3 9549 00356 2008